Getting Started with SAS® Enterprise Miner™ 12.3

SAS® Documentation

The correct bibliographic citation for this manual is as follows: SAS Institute Inc 2013. *Getting Started with SAS® Enterprise Miner™ 12.3*. Cary, NC: SAS Institute Inc.

Getting Started with SAS® Enterprise Miner™ 12.3

Copyright © 2013, SAS Institute Inc., Cary, NC, USA

ISBN 978-1-61290-774-1 (electronic book)
ISBN 978-1-61290-771-0

SAS Institute Inc., SAS Campus Drive, Cary, North Carolina 27513.

ISBN 978-1-61290-774-1
    Electronic book 1, July 2013

ISBN 978-1-61290-771-0
    Printing 1, July 2013

SAS® Publishing provides a complete selection of books and electronic products to help customers use SAS software to its fullest potential. For more information about our e-books, e-learning products, CDs, and hard-copy books, visit the SAS Publishing Web site at **support.sas.com/ publishing** or call 1-800-727-3228.

Contents

About This Book

Audience

This book is intended primarily for users who are new to SAS Enterprise Miner. The documentation assumes familiarity with graphical user interface (GUI) based software applications and basic, but not advanced, knowledge of data mining and statistical modeling principles. Although this knowledge is assumed, users who do not have this knowledge will still be able to complete the example that is described in this book end-to-end. In addition, SAS code is displayed in some result windows that are produced during the course of the example. However, SAS programming knowledge is not necessary to perform any task outlined in this book.

Recommended Reading

Here is the recommended reading list for this title:

*   The online Help for SAS Enterprise Miner

*   *Decision Trees for Business Intelligence and Data Mining: Using SAS Enterprise Miner*

*   *Introduction to Data Mining Using SAS Enterprise Miner*

*   *Customer Segmentation and Clustering Using SAS Enterprise Miner, Second Edition*

*   *Data Preparation for Analytics Using SAS*

*   *Predictive Modeling with SAS Enterprise Miner: Practical Solutions for Business Applications*

For a complete list of SAS books, go to support.sas.com/bookstore. If you have questions about which titles you need, please contact a SAS Book Sales Representative:

SAS Books
SAS Campus Drive
Cary, NC 27513-2414
Phone: 1-800-727-3228
Fax: 1-919-677-8166
E-mail: sasbook@sas.com
Web address: support.sas.com/bookstore

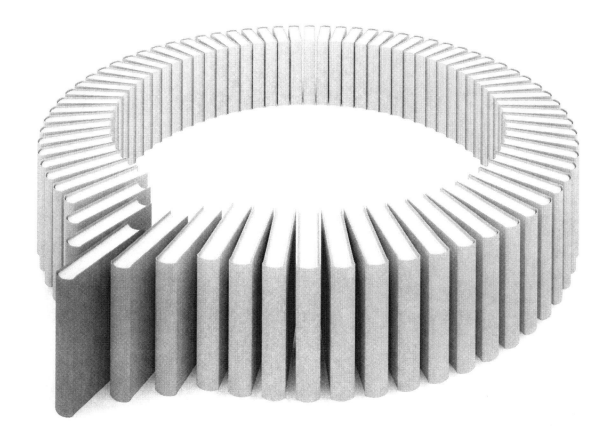

Gain Greater Insight into Your SAS® Software with SAS Books.

Discover all that you need on your journey to knowledge and empowerment.

support.sas.com/bookstore
for additional books and resources.

THE POWER TO KNOW®

Chapter 1
Introduction to SAS Enterprise Miner 12.3

What Is SAS Enterprise Miner?

SAS Enterprise Miner streamlines the data mining process to create highly accurate predictive and descriptive models based on analysis of vast amounts of data from across an enterprise. Data mining is applicable in a variety of industries and provides methodologies for such diverse business problems as fraud detection, householding, customer retention and attrition, database marketing, market segmentation, risk analysis, affinity analysis, customer satisfaction, bankruptcy prediction, and portfolio analysis.

In SAS Enterprise Miner, the data mining process has the following (SEMMA) steps:

- *Sample* the data by creating one or more data sets. The sample should be large enough to contain significant information, yet small enough to process. This step includes the use of data preparation tools for data import, merge, append, and filter, as well as statistical sampling techniques.

- *Explore* the data by searching for relationships, trends, and anomalies in order to gain understanding and ideas. This step includes the use of tools for statistical reporting and graphical exploration, variable selection methods, and variable clustering.

- *Modify* the data by creating, selecting, and transforming the variables to focus the model selection process. This step includes the use of tools for defining transformations, missing value handling, value recoding, and interactive binning.

- *Model* the data by using the analytical tools to train a statistical or machine learning model to reliably predict a desired outcome. This step includes the use of techniques such as linear and logistic regression, decision trees, neural networks, partial least squares, LARS and LASSO, nearest neighbor, and importing models defined by other users or even outside SAS Enterprise Miner.

- *Assess* the data by evaluating the usefulness and reliability of the findings from the data mining process. This step includes the use of tools for comparing models and computing new fit statistics, cutoff analysis, decision support, report generation, and score code management.

You might or might not include all of the SEMMA steps in an analysis, and it might be necessary to repeat one or more of the steps several times before you are satisfied with the results.

After you have completed the SEMMA steps, you can apply a scoring formula from one or more champion models to new data that might or might not contain the target variable. Scoring new data that is not available at the time of model training is the goal of most data mining problems.

Furthermore, advanced visualization tools enable you to quickly and easily examine large amounts of data in multidimensional histograms and to graphically compare modeling results.

Scoring new data that is not available at the time of model training is the goal of most data mining exercises. SAS Enterprise Miner includes tools for generating and testing complete score code for the entire process flow diagram as SAS Code, C code, and Java code, as well as tools for interactively scoring new data and examining the results. You can register your model to a SAS Metadata Server to share your results with users of applications such as SAS Enterprise Guide and SAS Data Integration Studio that can integrate the score code into reporting and production processes. SAS Model Manager complements the data mining process by providing a structure for managing projects through development, testing, and production environments and is fully integrated with SAS Enterprise Miner.

How Does SAS Enterprise Miner Work?

In SAS Enterprise Miner, the data mining process is driven by a process flow diagram that you create by dragging nodes from a toolbar that is organized by SEMMA categories and dropping them onto a diagram workspace.

Display 1.1 Example Process Flow Diagram

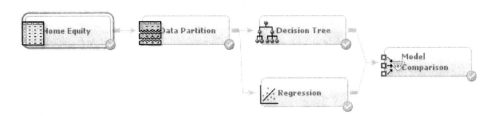

The graphical user interface (GUI) is designed in such a way that the business analyst who has little statistical expertise can navigate through the data mining methodology, and the quantitative expert can explore each node in depth to fine-tune the analytical process.

SAS Enterprise Miner automates the scoring process and supplies complete scoring code for all stages of model development in SAS, C, Java, and PMML. The scoring code can be deployed in a variety of real-time or batch environments within SAS, on the Web, or directly in relational databases.

Benefits of Using SAS Enterprise Miner

The benefits of using SAS Enterprise Miner include the following:

- **Support the entire data mining process with a broad set of tools.** Regardless of your data mining preference or skill level, SAS Enterprise Miner is flexible and addresses complex problems. Going from raw data to accurate, business-driven data mining models becomes a seamless process, enabling the statistical modeling group, business managers, and the IT department to collaborate more efficiently.

- **Build more models faster with an easy-to-use GUI.** The process flow diagram environment dramatically shortens model development time for both business analysts and statisticians. SAS Enterprise Miner includes an intuitive user interface that incorporates common design principles established for SAS software and additional navigation tools for moving easily around the workspace. The GUI can be tailored for all analysts' needs via flexible, interactive dialog boxes, code editors, and display settings.

- **Enhance accuracy of predictions.** Innovative algorithms enhance the stability and accuracy of predictions, which can be verified easily by visual model assessment and validation. Both analytical and business users enjoy a common, easy-to-interpret visual view of the data mining process. The process flow diagrams serve as self-documenting templates that can be updated easily or applied to new problems without starting over from scratch.

- **Surface business information and easily share results through the unique model repository.** Numerous integrated assessment features enable you to compare results of different modeling techniques in both statistical and business terms within a single, easy-to-interpret framework. SAS Enterprise Miner projects support the collaborative sharing of modeling results among quantitative analysts. Models can also be imported into the SAS Model Manager repository for sharing with scoring officers and independent model validation testers.

Accessibility Features of SAS Enterprise Miner 12.3

Overview of Accessibility Features

SAS Enterprise Miner 12.3 includes accessibility and compatibility features that improve the usability of the product for users with disabilities, with the exceptions noted below. These features are related to accessibility standards for electronic information technology that were adopted by the U.S. Government under Section 508 of the U.S. Rehabilitation Act of 1973, as amended.

SAS Enterprise Miner 12.3 conforms to accessibility standards for the Windows platform. For specific information about Windows accessibility features, refer to your operating system's help.

If you have questions or concerns about the accessibility of SAS products, send e-mail to `accessibility@sas.com`.

Exceptions to Standard Keyboard Controls

SAS Enterprise Miner 12.3 uses the same keyboard shortcuts as other Windows applications, with these exceptions:

- Instead of using the Windows standard, ALT+Spacebar, the system menu can be accessed by using these shortcuts:
  - primary window: Shift+F10+Spacebar
  - secondary window: Shift+F10+Down
- There is no keyboard equivalent for accessing the Explore window for a data source via the right-click pop-up menu. However, an alternate control is accessible from the **View** menu.
- There are no keyboard equivalents for these actions:
  - selecting a **SAS Server Directory** that is a subdirectory lower in the tree than the default folders in the Create New Project Wizard
  - selecting and editing the value of column attributes in the Data Source Wizard
  - maximizing or minimizing the Results window
  - accessing the Expression Builder in the Transform Variable node

Other Exceptions to Accessibility Standards

Other exceptions to the accessibility standards described in Section 508 of the U.S. Rehabilitation Act of 1973 include the following:

- On-screen indication of the current focus is not well-defined in some dialog boxes, in some menus, and in tables.
- High contrast color schemes are not universally inherited.
- SAS Enterprise Miner 12.3 is not fully accessible to assistive technologies:
  - Many controls are not read by JAWS, and the accessible properties of many controls are not surfaced to the Java Accessibility API.
  - Some content in the Data Source Wizard and Library Wizard is not accessible.

Getting to Know the Graphical User Interface

You use the SAS Enterprise Miner GUI to build a process flow diagram that controls your data mining project.

Display 1.2   *The SAS Enterprise Miner GUI*

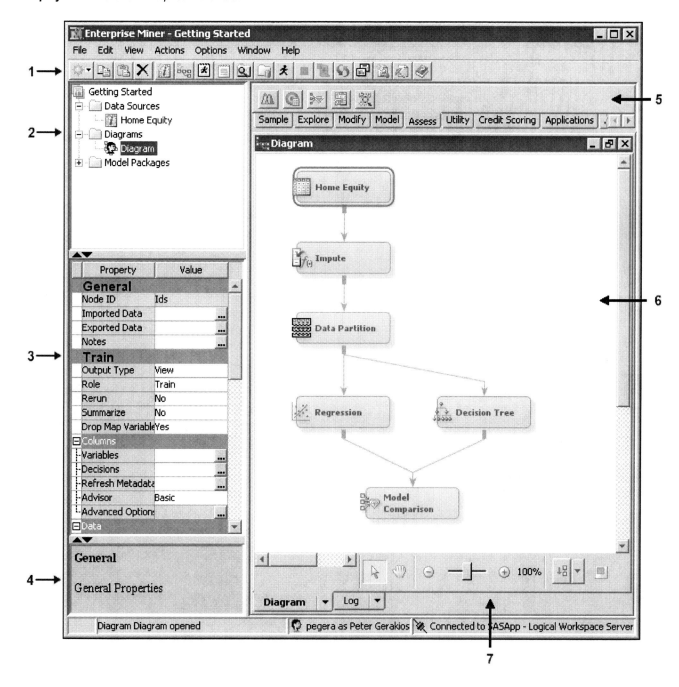

1. Toolbar Shortcut Buttons — Use the toolbar shortcut buttons to perform common computer functions and frequently used SAS Enterprise Miner operations. Move the mouse pointer over any shortcut button to see the text name. Click on a shortcut button to use it.

2. Project Panel — Use the Project Panel to manage and view data sources, diagrams, results, and project users.

3. Properties Panel — Use the Properties Panel to view and edit the settings of data sources, diagrams, nodes, and users.

4. Property Help Panel — The Property Help Panel displays a short description of any property that you select in the Properties Panel. Extended help can be found from the Help main menu.

5. Toolbar — The Toolbar is a graphic set of node icons that you use to build process flow diagrams in the Diagram Workspace. Drag a node icon into the Diagram Workspace to use it. The icon remains in place in the Toolbar, and the node in the Diagram Workspace is ready to be connected and configured for use in the process flow diagram.

6. Diagram Workspace — Use the Diagram Workspace to build, edit, run, and save process flow diagrams. In this workspace, you graphically build, order, sequence, and connect the nodes that you use to mine your data and generate reports.

7. Diagram Navigation Toolbar — Use the Diagram Navigation Toolbar to organize and navigate the process flow diagram.

Chapter 2

Learning by Example: Building and Running a Process Flow

About the Scenario in This Book

This book presents a basic data mining example that is intended to familiarize you with many features of SAS Enterprise Miner. In this example, you learn how to perform tasks that are required to build and run a process flow in order to solve a particular business problem. You should follow the chapters and the steps within the chapters in the order in which they are presented.

For the purpose of the scenario in this book, you are a data analyst at a national charitable organization. Your organization seeks to use the results of a previous postcard mail solicitation for donations to better target its next one. In particular, you want to determine which of the individuals in your mailing database have characteristics similar to those of your most profitable donors. By soliciting only these people, your organization can spend less money on the solicitation effort and more money on charitable concerns.

When you have finished building the process flow diagram as outlined in this example, the diagram will resemble the one shown below:

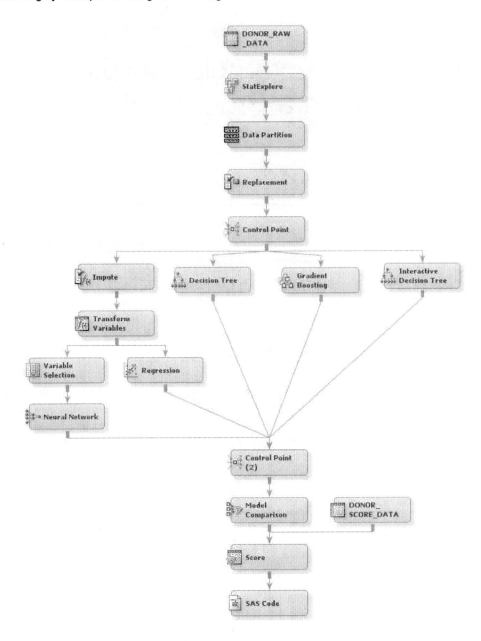

Prerequisites for This Example

In order to recreate this example, you must have access to SAS Enterprise Miner 12.3, either as a client/server application or as a complete installation on your local machine.

You must also have saved, on the SAS Enterprise Miner server machine, a copy of the sample data that is used in the example. You can download a .zip file that contains the data from `http://support.sas.com/documentation/onlinedoc/miner/`. Look for the item named **Example data for SAS Enterprise Miner 12.3**. For more information about the structure of the sample data, see Sample Data Reference on page 57.

Chapter 3
Set Up the Project

About the Tasks That You Will Perform

To perform these tasks, you need to have downloaded and unzipped the example data for SAS Enterprise Miner 12.3. If you have not, then see "Prerequisites for This Example" on page 8.

To set up the example project, you will perform the following tasks:

1. You will create a new SAS Enterprise Miner project.

2. You will define a new library that enables SAS Enterprise Miner to access the sample data.

3. You will define a new data source in the project, which is later used to import the sample data into a process flow.

4. You will create a new diagram within the project, and you will create the first node (for the input data source) in the process flow.

The steps in this example are written as if you were completing them in their entirety during one SAS Enterprise Miner session. However, you can easily complete the steps over multiple sessions. To return to the example project after you have closed and reopened SAS Enterprise Miner, click **Open Project** in the Welcome to Enterprise Miner window, and navigate to the saved project.

Create a New Project

In SAS Enterprise Miner, you store your work in projects. A project can contain multiple process flow diagrams and information that pertains to them.

TIP For organizational purposes, it is a good idea to create a separate project for each major data mining problem that you want to investigate.

To create the project that you will use in this example:

1. Open SAS Enterprise Miner.

2. In the Welcome to Enterprise Miner window, click **New Project**. The Create New Project Wizard opens.

3. Proceed through the steps below to complete the wizard. Contact your system administrator if you need to be granted directory access or if you are unsure about the details of your site's configuration.

   a. Select the logical workspace server to use. Click **Next**.

   b. Enter `Getting Started Charitable Giving Example` as the **Project Name**.

      The **SAS Server Directory** is the directory on the server machine in which SAS data sets and other files that are generated by the project will be stored. It is likely that your site is configured in such a way that the default path is appropriate for this example. Click **Next**.

   c. The **SAS Folder Location** is the directory on the server machine in which the project itself will be stored. It is likely that your site is configured in such a way that the default path is appropriate for the example project that you are about to create. Click **Next**.

      Note: If you complete this example over multiple sessions, then this is the location to which you should navigate after you select **Open Project** in the Welcome to Enterprise Miner window.

   d. Click **Finish**.

Create a Library

In order to access the sample data sets using SAS Enterprise Miner, you must create a SAS library to indicate to SAS the location in which they are stored. When you create a library, you give SAS a shortcut name and pointer to a storage location in your operating environment where you store SAS files.

To create a new SAS library for the sample data:

1. On the **File** menu, select **New** ⇨ **Library**. The Library Wizard opens.

2. Proceed through the steps below to complete the wizard. Contact your system administrator if you need to be granted directory access or if you are unsure about the details of your site's configuration.

   a. The **Create New Library** option button is automatically selected. Click **Next**.

   b. Enter `Donor` as the **Name**.

      Then enter the **Path** to the directory on the server machine that contains the sample data that you downloaded from the Web. For example, if the sample data is located on the desktop of the server machine (denoted by the C drive), then you could enter `C:\Users\<username>\Desktop`, where *<username>* is your user name on the server machine. Click **Next**.

   c. Click **Finish**.

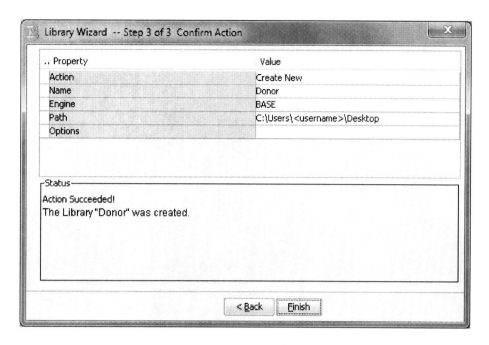

Create a Data Source

To use sample data that is stored in a SAS data set in a SAS Enterprise Miner project, you need to define a data source. In SAS Enterprise Miner, a data source stores the metadata of an input data set.

TIP You can also use input data saved in files (with extensions such as .jmp and .csv) that are not SAS data sets in a process flow. To import an external file into a process flow diagram, use the File Import node, which is located on the **Sample** tab on the Toolbar.

To create a new data source for the sample data:

1. On the **File** menu, select **New** ⇨ **Data Source**. The Data Source Wizard opens.

2. Proceed through the steps that are outlined in the wizard.

   a. **SAS Table** is automatically selected as the **Source**. Click **Next**.

   b. Enter `DONOR.DONOR_RAW_DATA` as the two-level filename of the **Table**. Click **Next**.

   c. Click **Next**.

   d. Select the **Advanced** option button. Click **Next**.

   e. Change the value of **Role** for the variables to match the description below. Then, click **Next**.

      • CONTROL_NUMBER should have the **Role** ID.

      • TARGET_B should have the **Role** Target.

      • TARGET_D should have the **Role** Rejected.

      • All other variables should have the **Role** Input.

      To change an attribute, click on the value of that attribute and select from the drop-down menu that appears.

Note: SAS Enterprise Miner automatically assigns the role **Target** to any variable whose name begins with the prefix TARGET_. For more information about the rules that SAS Enterprise Miner uses to automatically assign roles, see the SAS Enterprise Miner Help.

f. Select the **Yes** option button to indicate that you want to build models based on the values of decisions. Click **Next**.

- On the **Prior Probabilities** tab, select the **Yes** option button to indicate that you want to enter new prior probabilities. In the Adjusted Prior column of the table, enter **0.05** for Level 1 and **0.95** for Level 0.

Level	Count	Prior	Adjusted Prior
1	4843	0.25	0.05
0	14529	0.75	0.95

The values in the Prior column reflect the proportions of observations in the data set for which TARGET_B is equal to 1 and 0 (0.25 and 0.75, respectively). However, as the business analyst, you know that these proportions resulted from over-sampling of donors from the 97NK solicitation. In fact, you know that the true proportion of donors for the solicitation was closer to 0.05 than 0.25. For this reason, you adjust the prior probabilities.

- On the **Decision Weights** tab, the **Maximize** option button is automatically selected, which indicates that you want to maximize profit in this analysis.

Enter **14.5** as the Decision 1 weight for Level 1, **-0.5** as the Decision 1 weight for Level 0, and **0.0** as the Decision 2 weight for both levels. Click **Next**.

Level	DECISION1	DECISION2
1	14.5	0.0
0	-0.5	0.0

In this example, Decision 1 is the decision to mail a solicitation to an individual. Decision 2 is the decision to not mail a solicitation. If you mail a solicitation, and the individual does not respond, then your cost is $0.50 (the price of postage). However, if the individual does respond, then based on the previous solicitation, you expect to receive a $15.00 donation on average. Less the $0.50 postage cost, your organization expects $14.50 profit. If you do not mail a solicitation, you neither incur a cost nor expect a profit. These numbers are reflected in the decision weights that you entered in the table.

Click **Next**.

g. In the Data Source Wizard — Create Sample window, you decide whether to create a sample data set from the entire data source. This example uses the entire data set, so you need to select **No**. Click **Next**.

h. The **Role** of the data source is automatically selected as **Raw**. Click **Next**.

i. Click **Finish**.

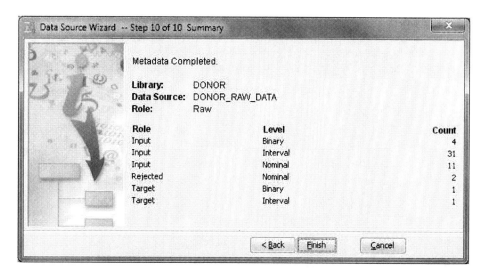

Create a Diagram and Add the Input Data Node

Now that you have created a project and defined the data source, you are ready to begin building a process flow diagram.

To create a process flow diagram and add the first node:

1. On the **File** menu, select **New** ⇨ **Diagram**.

2. Enter **Donations** as the **Diagram Name**, and click **OK**. An empty diagram opens in the Diagram Workspace.

3. Select the DONOR_RAW_DATA data source in the Project Panel. Drag it into the Diagram Workspace; this action creates the input data node.

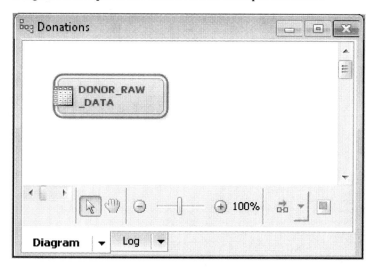

Chapter 4
Explore the Data and Replace Input Values

About the Tasks That You Will Perform

You have already set up the project and defined the input data source that you will use in this example. Now, you will import the data and perform the following tasks, which help you learn properties of the input data and prepare it for subsequent modeling:

1. You will explore the statistical properties of the variables in the input data set. The results that are generated in this step will give you an idea of which variables are most useful in predicting the target response (whether a person donates or not) in this data set.

2. You will partition the data into two data sets, a training data set and a validation data set. Such partitioning is common practice in data mining and enables you to develop a complete model that is not overfitted to a particular set of data.

3. You will specify how SAS Enterprise Miner should handle missing values of predictor variables.

   TIP It is always a good idea to plot the input data and to check it for missing values before you proceed to model building. Knowing the statistical properties of your input data is essential for building an accurate and robust predictive model.

Generate Descriptive Statistics

To use the StatExplore node to produce a statistical summary of the input data:

1. Select the **Explore** tab on the Toolbar.

2. Select the **StatExplore** node icon. Drag the node into the Diagram Workspace.

   TIP To determine which node an icon represents, position the mouse pointer over the icon and read the tooltip.

3. Connect the DONOR_RAW_DATA input data source node to the StatExplore node.

   To connect the two nodes, position the mouse pointer over the right edge of the input data source node until the pointer becomes a pencil. With the left mouse button held down, drag the pencil to the left edge of the **StatExplore** node. Then, release the mouse button. An arrow between the two nodes indicates a successful connection.

4. Select the **StatExplore** node. In the Properties Panel, scroll down to view the **Chi-Square Statistics** properties group. Click on the value of **Interval Variables** and select **Yes** from the drop-down menu that appears.

   Chi-square statistics are always computed for categorical variables. Changing the selection for interval variables causes SAS Enterprise Miner to distribute interval variables into five (by default) bins and compute chi-square statistics for the binned variables when you run the node.

   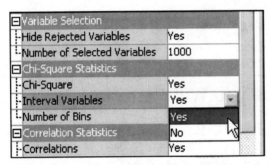

5. In the Diagram Workspace, right-click the **StatExplore** node, and select **Run** from the resulting menu. Click **Yes** in the Confirmation window that opens.

   When you run a node, all of the nodes preceding it in the process flow are also run in order, beginning with the first node that has changed since the flow was last run. If no nodes other than the one that you select have changed since the last run, then only the node that you select is run. You can watch the icons in the process flow diagram to monitor the status of execution.

   - Nodes that are outlined in green are currently running.

   - Nodes that are denoted with a check mark inside a green circle have successfully run.

   - Nodes that are outlined in red have failed to run due to errors.

   In this example, the DONOR_RAW_DATA input data node had not yet been run. Therefore, both nodes are run when you select to run the StatExplore node.

6. In the window that appears when processing completes, click **Results**. The Results window appears.

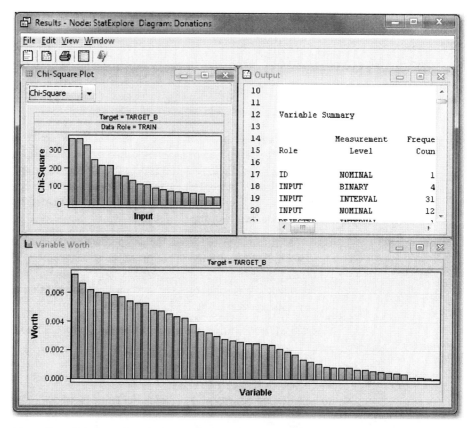

Note: Panels in Results windows might not have the same arrangement on your screen, due to window resizing. When the Results window is resized, SAS Enterprise Miner redistributes panels for optimal viewing.

The results window displays the following:

- a plot that orders the variables by their worth in predicting the target variable.

  Note: In the StatExplore node, SAS Enterprise Miner calculates variable worth using the Gini split worth statistic that would be generated by building a decision tree of depth 1. For detailed information about Gini split worth, see the SAS Enterprise Miner Help.

- the SAS output from the node.

- a plot that orders the top 20 variables by their chi-square statistics. You can also choose to view the top 20 variables ordered by their Cramer's V statistics on this plot.

  TIP In SAS Enterprise Miner, you can select graphs, tables, and rows within tables and select **Copy** from the right-click pop-up menu to copy these items for subsequent pasting in other applications such as Microsoft Word and Microsoft Excel.

7. Expand the Output window, and then scroll to the **Class Variable Summary Statistics** and the **Interval Variable Summary Statistics** sections of the output.

- Notice that there are two class variables and two interval variables for which there are missing values. Later in the example, you will impute values to use in the place of missing values for these variables.

- Notice that several variables have relatively large standard deviations. Later in the example, you will plot the data and explore transformations that can reduce the variances of these variables.

8. Close the Results window.

Partition the Data

In data mining, a strategy for assessing the quality of model generalization is to partition the data source. A portion of the data, called the *training data set*, is used for preliminary model fitting. The rest is reserved for empirical validation and is often split into two parts: validation data and test data. The *validation data set* is used to prevent a modeling node from overfitting the training data and to compare models. The *test data set* is used for a final assessment of the model.

Note: In SAS Enterprise Miner, the default data partitioning method for class target variables is to stratify on the target variable or variables. This method is appropriate for this sample data because there is a large number of non-donors in the input data relative to the number of donors. Stratifying ensures that both non-donors and donors are well-represented in the data partitions.

To use the Data Partition node to partition the input data into training and validation sets:

1. Select the **Sample** tab on the Toolbar.

2. Select the **Data Partition** node icon. Drag the node into the Diagram Workspace.

3. Connect the **StatExplore** node to the **Data Partition** node.

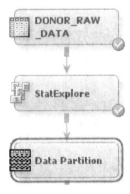

4. Select the **Data Partition** node. In the Properties Panel, scroll down to view the data set allocations in the Train properties.

    • Click on the value of **Training**, and enter **55.0**

    • Click on the value of **Validation**, and enter **45.0**

    • Click on the value of **Test**, and enter **0.0**

    These properties define the percentage of input data that is used in each type of mining data set. In this example, you use a training data set and a validation data set, but you do not use a test data set.

5. In the Diagram Workspace, right-click the **Data Partition** node, and select **Run** from the resulting menu. Click **Yes** in the Confirmation window that opens.

6. In the window that appears when processing completes, click **OK**.

Replace Missing Values

In this example, the variables SES and URBANICITY are class variables for which the value **?** denotes a missing value. Because a question mark does not denote a missing value in the terms that SAS defines a missing value (that is, a blank or a period), SAS Enterprise Miner sees it as an additional level of a class variable. However, the knowledge that these values are missing will be useful later in the model-building process.

To use the Replacement node to interactively specify that such observations of these variables are missing:

1. Select the **Modify** tab on the Toolbar.

2. Select the **Replacement** node icon. Drag the node into the Diagram Workspace.

3. Connect the **Data Partition** node to the **Replacement** node.

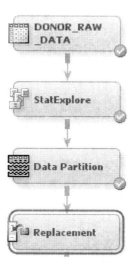

4. Select the **Replacement** node. In the Properties Panel, scroll down to view the Train properties.

   a. For interval variables, click on the value of **Default Limits Method**, and select **None** from the drop-down menu that appears. This selection indicates that no values of interval variables should be replaced. With the default selection, a particular range for the values of each interval variable would have been enforced. In this example, you do not want to enforce such a range.

      Note: In this data set, all missing interval variable values are correctly coded as SAS missing values (a blank or a period).

   b. For class variables, click on the ellipses that represent the value of **Replacement Editor**. The Replacement Editor opens.

      • Notice that SES and URBANICITY both have a level that contains observations with the value **?**. In the case of these two variables, this level represents observations with missing values. Enter **_MISSING_** as the **Replacement Value** for the two rows, as shown in the image below. This action enables SAS Enterprise Miner to see that the question marks indicate missing values for these two variables. Later, you will impute values for observations with missing values.

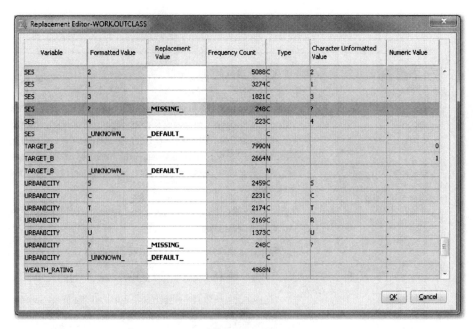

- Enter **_UNKNOWN_** as the **Replacement Value** for the level of DONOR_GENDER that has the value **A**. This value is the result of a data entry error, and you do not know whether the intention was to code it as an **F** or an **M**.

  Click **OK**.

5. In the Diagram Workspace, right-click the Replacement node, and select **Run** from the resulting menu. Click **Yes** in the Confirmation window that opens.

6. In the window that appears when processing completes, click **OK**.

Note: In the data that is exported from the Replacement node, a new variable is created for each variable that is replaced (in this example, SES, URBANICITY, and DONOR_GENDER). The original variable is not overwritten. Instead, the new variable has the same name as the original variable but is prefaced with REP_. The original version of each variable also exists in the exported data and has the role **Rejected**.

TIP To view the data that is exported by a node, click the ellipses that represent the value of the General property **Exported Data** in the Properties Panel. To view the exported variables, click **Properties** in the window that opens, and then view the **Variables** tab. Similarly, you can view the data that is imported and used by a node by clicking the ellipses that represent the value of the General property **Imported Data** in the Properties Panel.

Chapter 5
Build Decision Trees

About the Tasks That You Will Perform

Now that you have verified the input data, it is time to build predictive models. You perform the following tasks to model the input data using nonparametric decision trees:

1. You enable SAS Enterprise Miner to automatically train a full decision tree and to automatically prune the tree to an optimal size. When training the tree, you select split rules at each step to maximize the split decision logworth. Split decision logworth is a statistic that measures the effectiveness of a particular split decision at differentiating values of the target variable. For more information about logworth, see the SAS Enterprise Miner Help.

2. You interactively train a decision tree. At each step, you select from a list of candidate rules to define the split rule that you deem to be the best.

3. You use a Gradient Boosting node to generate a set of decision trees that form a single predictive model. Gradient boosting is a boosting approach that resamples the analysis data set several times to generate results that form a weighted average of the re-sampled data set.

Automatically Train and Prune a Decision Tree

Decision tree models are advantageous because they are conceptually easy to understand, yet they readily accommodate nonlinear associations between input variables and one or more target variables. They also handle missing values without the need for imputation. Therefore, you decide to first model the data using decision trees. You will compare decision tree models to other models later in the example.

However, before you add and run the Decision Tree node, you will add a Control Point node. The Control Point node is used to simplify a process flow diagram by reducing the number of connections between multiple interconnected nodes. By the end of this example, you will have created five different models of the input data set, and two Control Point nodes to connect these nodes. The first Control Point node, added here, will distribute the input data to each of these models. The second Control Point node will collect the models and send them to evaluation nodes.

To use the Control Point node:

1. Select the **Utility** tab on the Toolbar.

2. Select the **Control Point** node icon. Drag the node into the Diagram Workspace.

3. Connect the **Replacement** node to the **Control Point** node.

SAS Enterprise Miner enables you to build a decision tree in two ways: automatically and interactively. You will begin by letting SAS Enterprise Miner automatically train and prune a tree.

To use the **Decision Tree** node to automatically train and prune a decision tree:

1. Select the **Model** tab on the Toolbar.

2. Select the **Decision Tree** node icon. Drag the node into the Diagram Workspace.

3. Connect the **Control Point** node to the **Decision Tree** node.

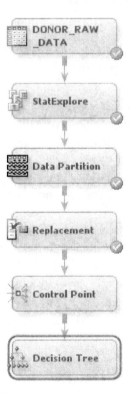

4. Select the **Decision Tree** node. In the Properties Panel, scroll down to view the **Train** properties:

    •   Click on the value of the **Maximum Depth** splitting rule property, and enter **10**. This specification enables SAS Enterprise Miner to train a tree that includes up to ten generations of the root node. The final tree in this example, however, will have fewer generations due to pruning.

    •   Click on the value of the **Leaf Size** node property, and enter **8**. This specification constrains the minimum number of training observations in any leaf to eight.

    •   Click on the value of the **Number of Surrogate Rules** node property, and enter **4**. This specification enables SAS Enterprise Miner to use up to four surrogate rules in each non-leaf node if the main splitting rule relies on an input whose value is missing.

Note: The **Assessment Measure** subtree property is automatically set to **Decision** because you defined a profit matrix in "Create a Data Source" on page 11. Accordingly, the Decision Tree node will build a tree that maximizes profit in the validation data.

5. In the Diagram Workspace, right-click the Decision Tree node, and select **Run** from the resulting menu. Click **Yes** in the Confirmation window that opens.

6. In the window that appears when processing completes, click **Results**. The Results window appears.

   a. On the **View** menu, select **Model** ⇨ **English Rules**. The English Rules window appears.

   b. Expand the English Rules window. This window contains the IF-THEN logic that distributes observations into each leaf node of the decision tree.

```
English Rules
 1     *-------------------------------------------------------------------*
 2       Node = 2
 3     *-------------------------------------------------------------------*
 4     if FREQUENCY_STATUS_97NK IS ONE OF: 3, 4
 5     then
 6       Tree Node Identifier   = 2
 7       Number of Observations = 3024.067307
 8       Predicted: TARGET_B=1  = 0.07
 9       Predicted: TARGET_B=0  = 0.93
10
11     *-------------------------------------------------------------------*
12       Node = 6
13     *-------------------------------------------------------------------*
14     if PEP_STAR IS ONE OF: 1
15     AND FREQUENCY_STATUS_97NK IS ONE OF: 1, 2 or MISSING
16     then
17       Tree Node Identifier   = 6
18       Number of Observations = 2989.1494063
19       Predicted: TARGET_B=1  = 0.05
20       Predicted: TARGET_B=0  = 0.95
21
22     *-------------------------------------------------------------------*
```

In the Output window, the **Tree Leaf Report** indicates that there are seven leaf nodes in this tree. For each leaf node, the following information is listed:

- node number

- number of training observations in the node

- percentage of training observations in the node with TARGET_B=1 (did donate), adjusted for prior probabilities

- percentage of training observations in the node with TARGET_B=0 (did not donate), adjusted for prior probabilities

This tree has been automatically pruned to an optimal size. Therefore, the node numbers that appear in the final tree are not sequential. In fact, they reflect the positions of the nodes in the full tree, before pruning.

7. Close the Results window.

Interactively Train a Decision Tree

To use the Decision Tree node to interactively train and prune a decision tree:

1. From the **Model** tab on the Toolbar, select the **Decision Tree** node icon. Drag the node into the Diagram Workspace.

2. In the Diagram Workspace, right-click the **Decision Tree** node, and select **Rename** from the resulting menu. Enter `Interactive Decision Tree` and then click **OK** in the window that opens.

3. Connect the Control Point node to the Interactive Decision Tree node.

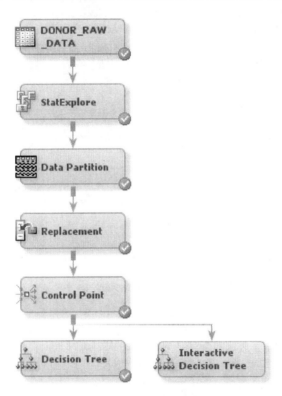

4. Select the **Interactive Decision Tree** node. In the Properties Panel, in the **Train** properties group, click on the ellipses that represent the value of **Interactive**. The Interactive Decision Tree window appears.

   a. Select the root node (at this point, the only node in the tree), and then from the **Action** menu, select **Split Node**. The Split Node window appears that lists the candidate splitting rules ranked by logworth (-Log(p)). The FREQUENCY_STATUS_97NK rule has the highest logworth. Ensure that this row is selected, and click **OK**.

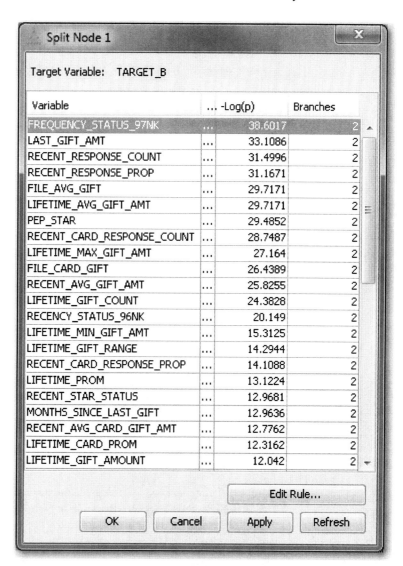

b. The tree now has two additional nodes. Select the lower left node (where FREQUENCY_STATUS_97NK is 3 or 4), and then from the **Action** menu, select **Split Node**. In the Split Node window that opens, select MONTHS_SINCE_LAST_GIFT, which ranks second in logworth, and click **Edit Rule** to manually specify the split point for this rule. The Interval Split Rule window appears.

Enter **8** as the **New split point**, and click **Add Branch**. Then, select Branch 3 (>= 8.5) and click **Remove Branch**. Click **OK**.

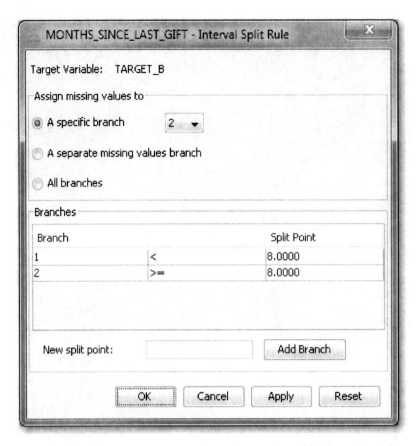

Ensure that MONTHS_SINCE_LAST_GIFT is selected in the Split Node window, and click **OK**.

c. Select the first generation node that you have not yet split (where FREQUENCY_STATUS_97NK is 1, 2, or Missing). From the **Action** menu, select **Split Node**. In the Split Node window that opens, ensure that PEP_STAR is selected, and click **OK**.

The tree now has seven nodes, four of which are leaf nodes. The nodes are colored from light to dark, corresponding to low to high percentages of correctly classified observations.

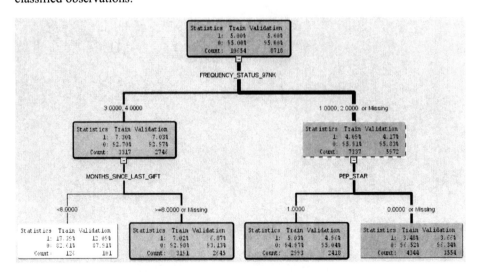

d. Select the lower right node (where FREQUENCY_STATUS_97NK is 1, 2, or Missing and PEP_STAR is 0 or Missing). From the **Action** menu, select **Train Node**. This selection causes SAS Enterprise Miner to continue adding generations of this node until a stopping criterion is met. For more information about stopping criteria for decision trees, see the SAS Enterprise Miner Help.

> *Note:* In the Interactive Decision Tree window, you can prune decision trees. However, in this example, you will leave the tree in its current state.

    e. Close the Interactive Decision Tree window.

Create a Gradient Boosting Model of the Data

The Gradient Boosting node uses a partitioning algorithm to search for an optimal partition of the data for a single target variable. Gradient boosting is an approach that resamples the analysis data several times to generate results that form a weighted average of the resampled data set. Tree boosting creates a series of decision trees that form a single predictive model.

Like decision trees, boosting makes no assumptions about the distribution of the data. Boosting is less prone to overfit the data than a single decision tree. If a decision tree fits the data fairly well, then boosting often improves the fit. For more information about the Gradient Boosting node, see the SAS Enterprise Miner help documentation.

To create a gradient boosting model of the data:

1. Select the **Model** tab on the Toolbar.

2. Select the **Gradient Boosting** node icon. Drag the node into the Diagram Workspace.

3. Connect the **Control Point** node to the **Gradient Boosting** node.

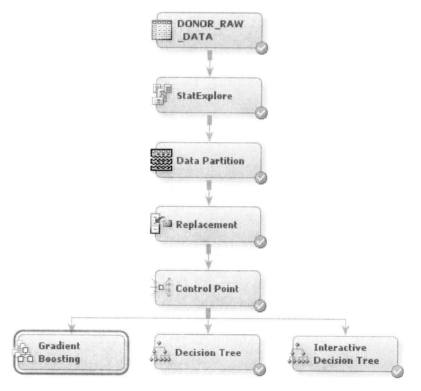

4. Select the **Gradient Boosting** node. In the Properties Panel, set the following properties:

- Click on the value for the **Maximum Depth** property, in the **Splitting Rule** subgroup, and enter **10**. This property determines the number of generations in each decision tree created by the Gradient Boosting node.

- Click on the value for the **Number of Surrogate Rules** property, in the **Node** subgroup, and enter **2**. Surrogate rules are backup rules that are used in the event of missing data. For example, if your primary splitting rule sorts donors based on their ZIP codes, then a reasonable surrogate rule would sort based on the donor's city of residence.

5. In the Diagram Workspace, right-click the Gradient Boosting node, and select **Run** from the resulting menu. Click **Yes** in the Confirmation window that opens.

6. In the Run Status window, select **OK**.

Chapter 6
Impute and Transform, Build Neural Networks, and Build a Regression Model

About the Tasks That You Will Perform

You have just modeled the input data using decision trees, which are nonparametric. As part of your analysis, you now perform the following tasks in order to also model the data using parametric methods:

1. You impute values to use as replacements for missing values that are in the input data. Regressions and neural networks would otherwise ignore missing values, which would decrease the amount of data that you use in the models and lower their predictive power.

2. You transform input variables to make the usual assumptions of regression more appropriate for the input data.

3. You model the input data using logistic regression, a statistical method with which your management is familiar.

4. You model the input data using neural networks, which are more flexible than logistic regression (and more complicated).

Impute Missing Values

For decision trees, missing values are not problematic. Surrogate splitting rules enable you to use the values of other input variables to perform a split for observations with missing values. In SAS Enterprise Miner, however, models such as regressions and neural networks ignore altogether observations that contain missing values, which reduces the size of the training data set. Less training data can substantially weaken the predictive power of these models. To overcome this obstacle of missing data, you can impute missing values before you fit the models.

TIP It is a particularly good idea to impute missing values before fitting a model that ignores observations with missing values if you plan to compare those models with a decision tree. Model comparison is most appropriate between models that are fit with the same set of observations.

To use the Impute node to impute missing values:

1. Select the **Modify** tab on the Toolbar.

2. Select the **Impute** node icon. Drag the node into the Diagram Workspace.

3. Connect the **Control Point** node to the **Impute** node.

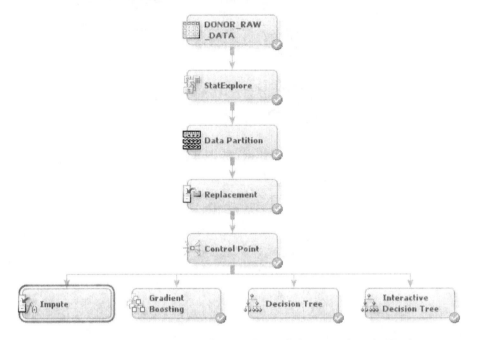

4. Select the **Impute** node. In the Properties Panel, scroll down to view the **Train** properties:

   - For class variables, click on the value of **Default Input Method** and select **Tree Surrogate** from the drop-down menu that appears.

   - For interval variables, click on the value of **Default Input Method** and select **Median** from the drop-down menu that appears.

   The default input method specifies which default statistic to use to impute missing values. In this example, the values of missing interval variables are replaced by the median of the nonmissing values. This statistic is less sensitive to extreme values than the mean or midrange and is therefore useful for imputation of missing values from skewed distributions. The values of missing class variables, in this example, are imputed using predicted values from a decision tree. For each class variable, SAS Enterprise Miner builds a decision tree (in this case, potentially using surrogate splitting rules) with that variable as the target and the other input variables as predictors.

5. In the Diagram Workspace, right-click the **Impute** node, and select **Run** from the resulting menu. Click **Yes** in the Confirmation window that opens.

6. In the window that appears when processing completes, click **OK**.

Note: In the data that is exported from the Impute node, a new variable is created for each variable for which missing values are imputed. The original variable is not overwritten. Instead, the new variable has the same name as the original variable but is prefaced with IMP_. The original version of each variable also exists in the exported data and has the role `Rejected`. In this example, SES and URBANICITY have values replaced and then imputed. Therefore, in addition to the original version, each of these variables has a version in the exported data that is prefaced by IMP_REP_.

Transform Variables

Sometimes, input data is more informative on a scale other than that from which it was originally collected. For example, variable transformations can be used to stabilize variance, remove nonlinearity, improve additivity, and counter non-normality. Therefore, for many models, transformations of the input data (either dependent or independent variables) can lead to a better model fit. These transformations can be functions of either a single variable or of more than one variable.

To use the Transform Variables node to make variables better suited for logistic regression models and neural networks:

1. From the **Modify** tab on the Toolbar, select the **Transform Variables** node icon. Drag the node into the Diagram Workspace.

2. Connect the **Impute** node to the **Transform Variables** node.

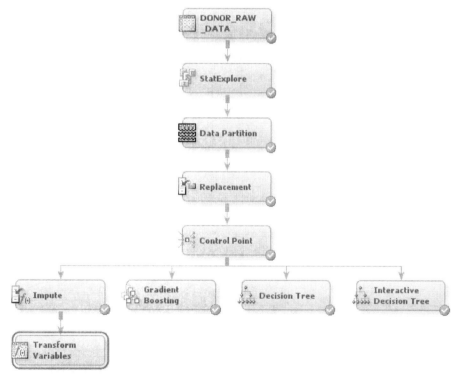

   TIP To align a process flow diagram vertically, as in the image above, right-click anywhere in the Diagram Workspace, and select **Layout** ⇨ **Vertically** from the resulting menu.

3. Select the **Transform Variables** node. In the Properties Panel, scroll down to view the **Train** properties, and click on the ellipses that represent the value of **Formulas**. The Formulas window appears.

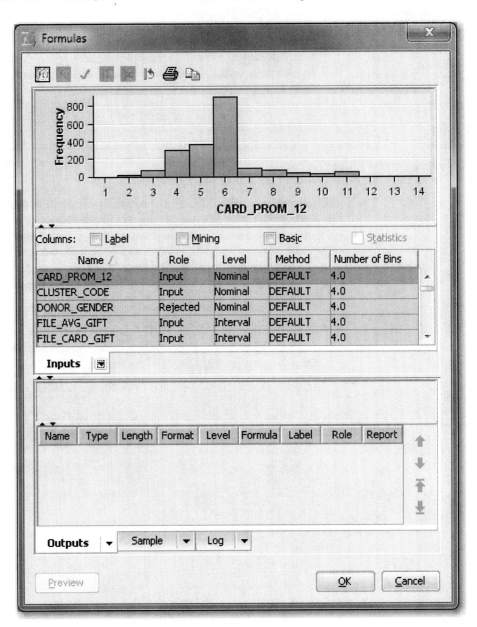

a. In the variables table, click the **Role** column heading to sort the variables in ascending order by their role.

b. You can select any row in the variable table to display the histogram of a variable in the panel above. Look at the histograms for all variables that have the role `Input`. Notice that several variables have skewed distributions.

c. Click **OK** to close the Formulas window.

4. In the Properties Panel, scroll down to view the **Train** properties, and click on the ellipses that represent the value of **Variables**. The Variables — Trans window appears.

a. The common log transformation is often used to control skewness. Select the transformation **Method** for the following interval variables and select **Log 10** from the drop-down menu that appears:

   • FILE_AVG_GIFT

   • LAST_GIFT_AMT

- LIFETIME_AVG_GIFT_AMT

- LIFETIME_GIFT_AMOUNT

    TIP You can hold down the CTRL key to select multiple rows. Then, when you select a new **Method** for one of the selected variables, the new method will apply to all of the selected variables.

    b. Select the transformation **Method** for the following interval variables and select **Optimal Binning** from the drop-down menu that appears:

    - LIFETIME_CARD_PROM

    - LIFETIME_GIFT_COUNT

    - MEDIAN_HOME_VALUE

    - MEDIAN_HOUSEHOLD_INCOME

    - PER_CAPITA_INCOME

    - RECENT_RESPONSE_PROP

    - RECENT_STAR_STATUS

    The optimal binning transformation is useful when there is a nonlinear relationship between an input variable and the target. For more information about this transformation, see the SAS Enterprise Miner Help.

    c. Click **OK**.

5. In the Diagram Workspace, right-click the Transform Variables node, and select **Run** from the resulting menu. Click **Yes** in the Confirmation window that opens.

6. In the window that appears when processing completes, click **OK**.

Note: In the data that is exported from the Transform Variables node, a new variable is created for each variable that is transformed. The original variable is not overwritten. Instead, the new variable has the same name as the original variable but is prefaced with an identifier of the transformation. For example, variables to which the log transformation have been applied are prefaced with LOG_, and variables to which the optimal binning transformation have been applied are prefaced with OPT_. The original version of each variable also exists in the exported data and has the role `Rejected`.

Analyze with a Logistic Regression Model

As part of your analysis, you want to include some parametric models for comparison with the decision trees that you built in Chapter 5, "Build Decision Trees," on page 21. Because it is familiar to the management of your organization, you have decided to include a logistic regression as one of the parametric models.

To use the Regression node to fit a logistic regression model:

1. Select the **Model** tab on the Toolbar.

2. Select the **Regression** node icon. Drag the node into the Diagram Workspace.

3. Connect the **Transform Variables** node to the **Regression** node.

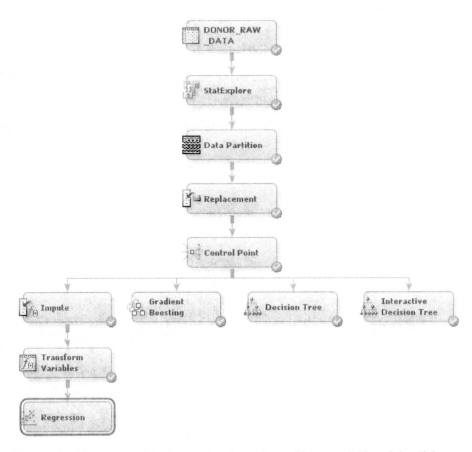

4. To examine histograms of the imputed and transformed input variables, right-click the Regression node and select **Update**. In the diagram workspace, select the Regression node. In the Properties Panel, scroll down to view the Train properties, and click on the ellipses that represent the value of **Variables**. The Variables — Reg window appears.

   a. Select all variables that have the prefix LG10_. Click **Explore**. The Explore window appears.

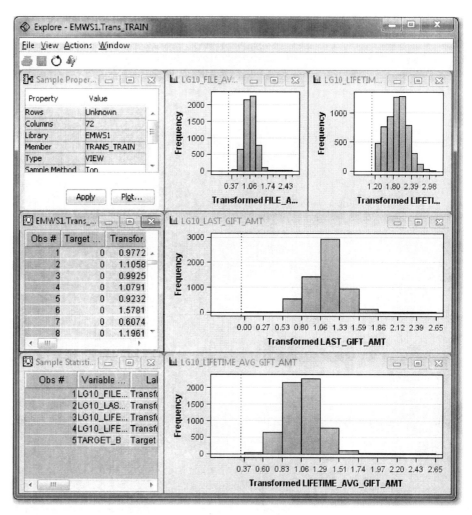

You can select a bar in any histogram, and the observations that are in that bucket are highlighted in the EMWS.Trans_TRAIN data set window and in the other histograms. Close the Explore window to return to the Variables — Reg window.

b. (Optional) You can explore the histograms of other input variables.

c. Close the Variables — Reg window.

5. In the Properties Panel, scroll down to view the Train properties. Click on the **Selection Model** property in the **Model Selection** subgroup, and select **Stepwise** from the drop-down menu that appears. This specification causes SAS Enterprise Miner to use stepwise variable selection to build the logistic regression model.

   Note: The Regression node automatically performs logistic regression if the target variable is a class variable that takes one of two values. If the target variable is a continuous variable, then the Regression node performs linear regression.

6. In the Diagram Workspace, right-click the Regression node, and select **Run** from the resulting menu. Click **Yes** in the Confirmation window that opens.

7. In the window that appears when processing completes, click **Results**. The Results window appears.

8. Maximize the Output window. This window details the variable selection process. Lines 401 – 424 list a summary of the steps that were taken.

9. Minimize the Output window and maximize the Score Rankings Overlay window. From the drop-down menu, select **Cumulative Total Expected Profit**.

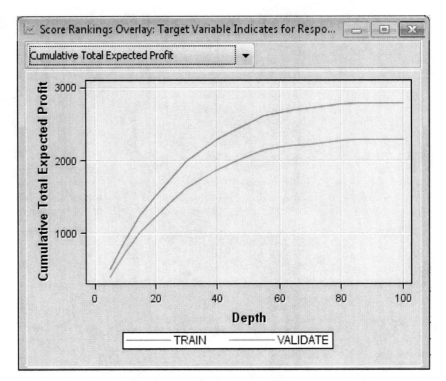

The data that is used to construct this plot is ordered by expected profit. For this example, you have defined a profit matrix. Therefore, expected profit is a function of both the probability of donation for an individual and the profit associated with the corresponding outcome. A value is computed for each decision from the sum of the decision matrix values multiplied by the classification probabilities and minus any defined cost. The decision with the greatest value is selected, and the value of that selected decision for each observation is used to compute overall profit measures.

The plot represents the cumulative total expected profit that results from soliciting the best *n*% of the individuals (as determined by expected profit) on your mailing list. For example, if you were to solicit the best 40% of the individuals, the total expected profit from the validation data would be approximately $1850. If you were to solicit everyone on the list, then based on the validation data, you could expect approximately $2250 profit on the campaign.

10. Close the Results window.

Analyze with a Neural Network Model

Neural networks are a class of parametric models that can accommodate a wider variety of nonlinear relationships between a set of predictors and a target variable than can logistic regression. Building a neural network model involves two main phases. First, you must define the network configuration. You can think of this step as defining the structure of the model that you want to use. Then, you iteratively train the model.

A neural network model will be more complicated to explain to the management of your organization than a regression or a decision tree. However, you know that the management would prefer a stronger predictive model, even if it is more complicated. So, you decide to run a neural network model, which you will compare to the other models later in the example.

Because neural networks are so flexible, SAS Enterprise Miner has two nodes that fit neural network models: the Neural Network node and the AutoNeural node. The Neural Network node trains a specific neural network configuration; this node is best used when you know a lot about the structure of the model that you want to define. The AutoNeural node searches over several network configurations to find one that best describes the relationship in a data set and then trains that network.

This example does not use the AutoNeural node. However, you are encouraged to explore the features of this node on your own.

Before creating a neural network, you will reduce the number of input variables with the Variable Selection node. Performing variable selection reduces the number of input variables and saves computer resources. To use the Variable Selection node to reduce the number of input variables that are used in a neural network:

1. Select the **Explore** tab on the Toolbar.

2. Select the **Variable Selection** node icon. Drag the node into the Diagram Workspace.

3. Connect the **Transform Variables** node to the **Variable Selection** node.

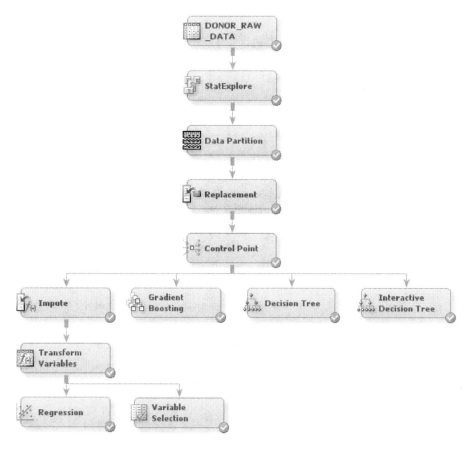

4. In the Diagram Workspace, right-click the Variable Selection node, and select **Run** from the resulting menu. Click **Yes** in the Confirmation window that opens.

5. In the window that appears when processing completes, click **Results**. The Results window appears.

6. Expand the Variable Selection window.

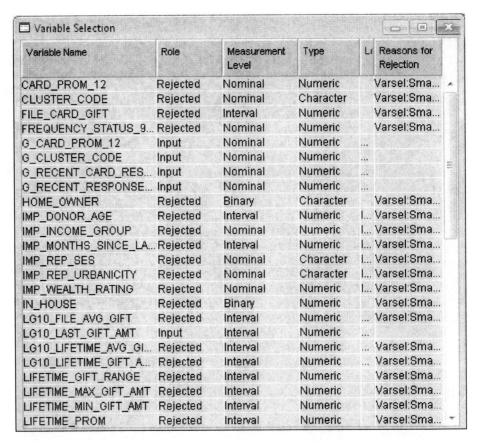

Variable Name	Role	Measurement Level	Type	Lε	Reasons for Rejection
CARD_PROM_12	Rejected	Nominal	Numeric		Varsel:Sma...
CLUSTER_CODE	Rejected	Nominal	Character		Varsel:Sma...
FILE_CARD_GIFT	Rejected	Interval	Numeric		Varsel:Sma...
FREQUENCY_STATUS_9...	Rejected	Nominal	Numeric		Varsel:Sma...
G_CARD_PROM_12	Input	Nominal	Numeric	...	
G_CLUSTER_CODE	Input	Nominal	Numeric	...	
G_RECENT_CARD_RES...	Input	Nominal	Numeric	...	
G_RECENT_RESPONSE...	Input	Nominal	Numeric	...	
HOME_OWNER	Rejected	Binary	Character		Varsel:Sma...
IMP_DONOR_AGE	Rejected	Interval	Numeric	I...	Varsel:Sma...
IMP_INCOME_GROUP	Rejected	Nominal	Numeric	I...	Varsel:Sma...
IMP_MONTHS_SINCE_LA...	Rejected	Interval	Numeric	I...	Varsel:Sma...
IMP_REP_SES	Rejected	Nominal	Character	I...	Varsel:Sma...
IMP_REP_URBANICITY	Rejected	Nominal	Character	I...	Varsel:Sma...
IMP_WEALTH_RATING	Rejected	Nominal	Numeric	I...	Varsel:Sma...
IN_HOUSE	Rejected	Binary	Numeric		Varsel:Sma...
LG10_FILE_AVG_GIFT	Rejected	Interval	Numeric	...	Varsel:Sma...
LG10_LAST_GIFT_AMT	Input	Interval	Numeric	...	
LG10_LIFETIME_AVG_GI...	Rejected	Interval	Numeric	...	Varsel:Sma...
LG10_LIFETIME_GIFT_A...	Rejected	Interval	Numeric	...	Varsel:Sma...
LIFETIME_GIFT_RANGE	Rejected	Interval	Numeric		Varsel:Sma...
LIFETIME_MAX_GIFT_AMT	Rejected	Interval	Numeric		Varsel:Sma...
LIFETIME_MIN_GIFT_AMT	Rejected	Interval	Numeric		Varsel:Sma...
LIFETIME_PROM	Rejected	Interval	Numeric		Varsel:Sma...

Examine the table to see which variables were selected. The role for variables that were not selected has been changed to **Rejected**. Close the Results window.

Note: In this example, for variable selection, a forward stepwise least squares regression method was used that maximizes the model R-square value. For more information about this method, see the SAS Enterprise Miner Help.

7. Close the Results window.

The input data is now ready to be modeled with a neural network. To use the Neural Network node to train a specific neural network configuration:

1. From the **Model** tab on the Toolbar, select the **Neural Network** node icon. Drag the node into the Diagram Workspace.

2. Connect the **Variable Selection** node to the **Neural Network** node.

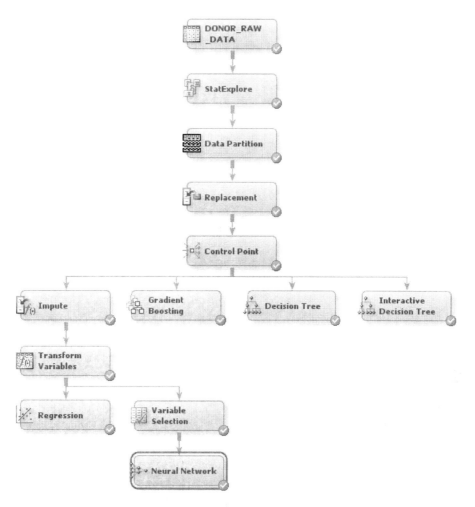

3. Select the **Neural Network** node. In the Properties Panel, scroll down to view the **Train** properties, and click on the ellipses that represent the value of **Network**. The Network window appears. For more information about neural networks, connections, and hidden units, see the Neural Network Node: Reference documentation in SAS Enterprise Miner help.

   Change the following properties:

   • Click on the value of **Direct Connection** and select **Yes** from the drop-down menu that appears. This selection enables the network to have connections directly between the inputs and the outputs in addition to connections via the hidden units.

   • Click on the value of **Number of Hidden Units** and enter **5**. This example trains a multilayer perceptron neural network with five units on the hidden layer.

   Click **OK**.

4. In the Diagram Workspace, right-click the Neural Network node, and select **Run** from the resulting menu. Click **Yes** in the Confirmation window that opens.

5. In the window that appears when processing completes, click **Results**. The Results window appears. Maximize the Score Rankings Overlay window. From the drop-down menu, select **Cumulative Total Expected Profit**.

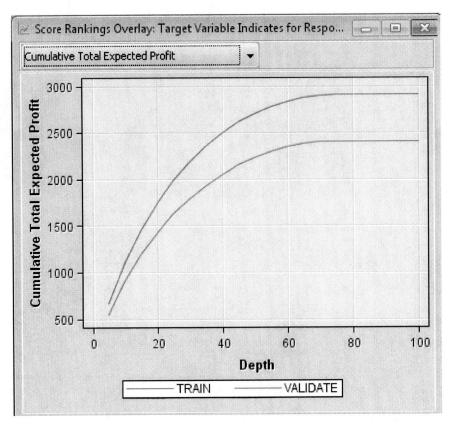

Compare these results to those from the Regression node. According to this model, if you were to solicit the best 40% of the individuals, the total expected profit from the validation data would be approximately $2050. If you were to solicit everyone on the list, then based on the validation data, you could expect approximately $2400 profit on the campaign.

6. Close the Results window.

Chapter 7
Compare Models and Score New Data

About the Tasks That You Will Perform

Now that you have five candidate models to use to predict the best target individuals for your mail solicitation, you can compare them to determine a champion model that you will use to score new data. You perform the following tasks in order to determine which of the individuals in your organization's database to solicit:

1. You compare models and select a champion model that, according to an evaluation criterion, performs best in the validation data.

2. You create a new data source for a data set that contains scoring data that has not been used to build any of the models thus far in the process flow and that does not include values of the target variable. You score this data using the champion model.

3. You write SAS code to output, based on the scored data, a list of the top potential donors according to the probability of donating and the profitability matrix that you defined in "Create a Data Source" on page 11.

Compare Models

To use the Model Comparison node to compare the models that you have built in this example and to select one as the champion model:

1. Select the **Utility** tab on the Toolbar.

2. Select the **Control Point** node icon. Drag the node into the Diagram Workspace.

3. Connect all five model nodes to the **Control Point** node.

   TIP Control Point nodes enable you to better organize your process flow diagram. These nodes do not perform calculations; they simply pass data from preceding nodes to subsequent nodes.

4. Select the **Assess** tab on the Toolbar.

5. Select the **Model Comparison** node icon. Drag the node into the Diagram Workspace.

6. Connect the **Control Point** node to the **Model Comparison** node.

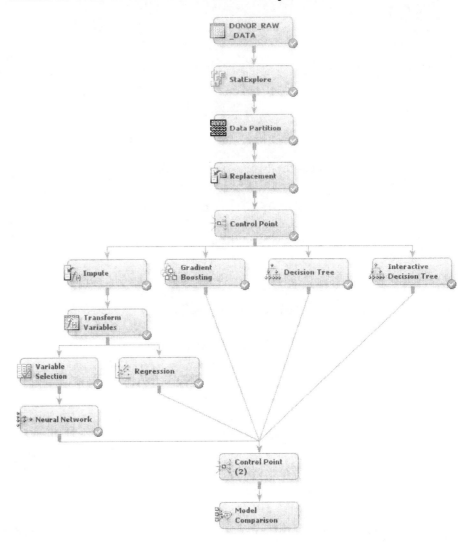

7. In the Diagram Workspace, right-click the **Model Comparison** node, and select **Run** from the resulting menu. Click **Yes** in the Confirmation window that opens.

8. In the window that appears when processing completes, click **Results**. The Results window appears.

9. In the Fit Statistics window, notice that the regression model was selected as the champion model. The champion model has the value **Y** in the Selected Model column in the Fit Statistics window.

   In the model selection node, SAS Enterprise Miner selects the champion model based on the value of a single statistic. You can specify which statistic to use for selection in the node Properties Panel. Because you did not change the value of this property, the default statistic was used, which (because a profit matrix is defined) is the average profit in the validation data.

10. Close the Results window.

Score New Data

To create a new data source for the sample data that you will score:

1. On the **File** menu, select **New** ⇨ **Data Source**. The Data Source Wizard opens.

2. Proceed through the steps that are outlined in the wizard.

   a. **SAS Table** is automatically selected as the **Source**. Click **Next**.

   b. Enter `DONOR.DONOR_SCORE_DATA` as the two-level filename of the **Table**. Click **Next**.

   c. Click **Next**.

   d. The **Basic** option is automatically selected. Click **Next**.

   e. Click **Next**.

   f. The **No** option is automatically selected. Click **Next**.

   g. Select **Score** as the **Role** of the data source, and click **Next**.

   h. Click **Finish**.

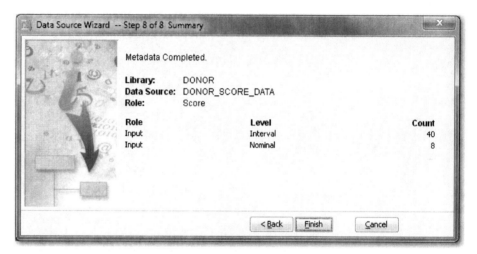

3. Select the DONOR_SCORE_DATA data source in the Project Panel. Drag it into the Diagram Workspace.

To use the Score node to score new data:

1. Select the **Assess** tab on the Toolbar.

2. Select the **Score** node icon. Drag the node into the Diagram Workspace.

3. Connect the **Model Comparison** node and the DONOR_SCORE_DATA data source node to the Score node.

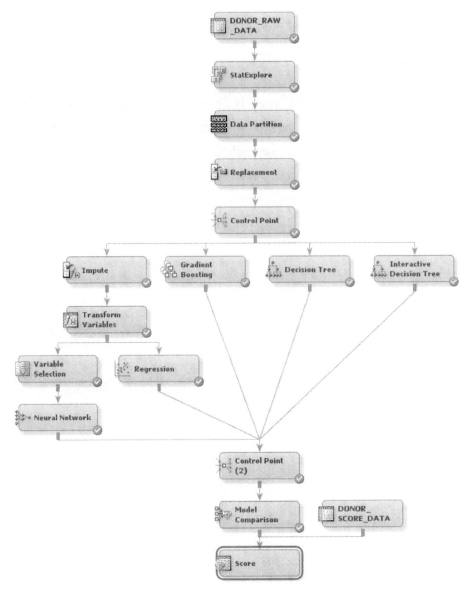

4. In the Diagram Workspace, right-click the **Score** node, and select **Run** from the resulting menu. Click **Yes** in the Confirmation window that opens.

5. In the window that appears when processing completes, click **OK**. The scoring code has been generated, and the scoring data has been scored.

   Note: The SAS scoring code is viewable in the node results.

Create a Sorted List of Potential Donors

To use the SAS Code node to create a list of best potential donors:

1. Select the **Utility** tab on the Toolbar.

2. Select the **SAS Code** node icon. Drag the node into the Diagram Workspace.

3. Connect the **Score** node to the **SAS Code** node.

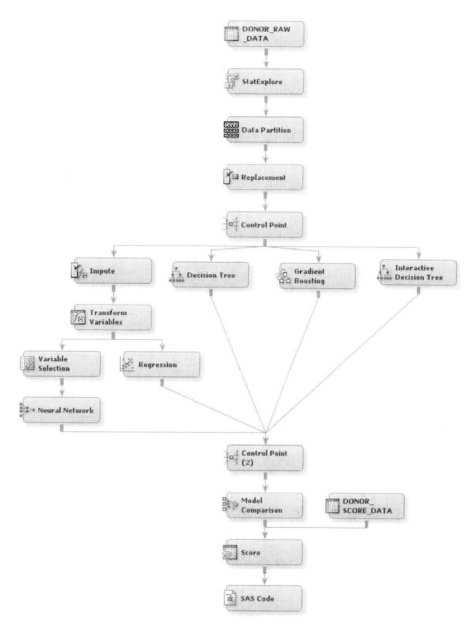

4. Select the **SAS Code** node. In the Properties Panel, scroll down to view the **Train** properties, and click on the ellipses that represent the value of **Code Editor**. The Training Code — Code Node window appears.

    a. In the **Training Code** box, enter the following SAS code, where *<LIBREF>* is the libref of the diagram:

    ```
    proc sort data= <LIBREF>.Score_SCORE out= bestlist;
    by descending ep_target_b;
    run;

    proc print data= bestlist;
    var control_number ep_target_b;
    run;
    ```

    > **TIP** To determine the libref of the diagram, select the diagram in the Project Panel and then view the diagram ID in the Properties Panel. The diagram ID is the libref that you should use.

    b. From the **File** menu, select **Save All**. Then, close the window.

5. In the Diagram Workspace, right-click the SAS Code node, and select **Run** from the resulting menu. Click **Yes** in the Confirmation window that opens. In the Run Status window, click **Results**. The Results window appears in which you can view the scored list of potential donors sorted by expected profit.

6. Close the Results window.

Appendix 1
SAS Enterprise Miner Node Reference

About Nodes

Most SAS Enterprise Miner nodes are organized on tabs according to the SEMMA data mining methodology. There are also tabs for Credit Scoring and Utility node groups. Use the Credit Scoring nodes to score data models and to create freestanding code. Use the Utility nodes to submit SAS programming statements and to define control points in the process flow diagram.

Note: The Credit Scoring nodes do not appear in all installed versions of SAS Enterprise Miner. For more information about the Credit Scoring nodes, see the SAS Enterprise Miner Credit Scoring Help.

Table A1.1 Sample Nodes

Node Name	Description
Append	Use the Append node to append data sets that are exported by two different paths in a single process flow diagram. The Append node can also append train, validation, and test data sets into a new training data set.
Data Partition	Use the Data Partition node to partition an input data set into a training, test, and validation data set. The training data set is used for preliminary model fitting. The validation data set is used to monitor and tune the free model parameters during estimation. It is also used for model assessment. The test data set is an additional holdout data set that you can use for model assessment.
File Import	Use the File Import node to convert selected external flat files, spreadsheets, and database tables into a format that SAS Enterprise Miner recognizes as a data source and can use in data mining process flow diagrams.
Filter	Use the Filter node to create and apply filters to the input data. You can use filters to exclude certain observations, such as extreme outliers and errant data that you do not want to include in a mining analysis.

Node Name	Description
Input Data	The Input Data node represents the data source that you choose for a mining analysis. It provides details (metadata) about the variables in the data source that you want to use.
Merge	Use the Merge node to merge observations from two or more data sets into a single observation in a new data set.
Sample	Use the Sample node to extract a simple random sample, n^{th}-observation sample, stratified sample, first-n sample, or cluster sample of an input data set. Sampling is recommended for extremely large databases because it can significantly decrease model training time. If the random sample sufficiently represents the source data set, then data relationships that SAS Enterprise Miner finds in the sample can be applied to the complete source data set.
Time Series	Use the Time Series node to convert transactional data to time series data to perform seasonal and trend analysis. Transactional data is timestamped data that is collected over time at no particular frequency. By contrast, time series data is timestamped data that is collected over time at a specific frequency.

Table A1.2 *Explore Nodes*

Node Name	Description
Association	Use the Association node to identify association and sequence relationships within the data. For example, "If a customer buys cat food, how likely is the customer to also buy cat litter?" In the case of sequence discovery, this question could be extended and posed as, "If a customer buys cat food today, how likely is the customer to buy cat litter within the next week?"
Cluster	Use the Cluster node to perform observation clustering, which can be used to segment databases. Clustering places objects into groups or clusters suggested by the data. The objects in each cluster tend to be similar to each other in some sense, and objects in different clusters tend to be dissimilar.
DMDB (Data Mining Database)	The DMDB node creates a data mining database that provides summary statistics and factor-level information for class and interval variables in the imported data set. Improvements to SAS Enterprise Miner have eliminated the previous need to use the DMDB node to optimize the performance of nodes. However, the DMDB database can still provide quick summary statistics for class and interval variables at a given point in a process flow diagram.
Graph Explore	The Graph Explore node is an advanced visualization tool that enables you to interactively explore large volumes of data to uncover patterns and trends and to reveal extreme values in the database. You can analyze univariate distributions, investigate multivariate distributions, create scatter and box plots, constellation and 3-D charts, and so on.

Node Name	Description
Market Basket	The Market Basket node performs association rule mining over transaction data in conjunction with item taxonomy. Market basket analysis uses the information from the transaction data to give you insight (for example, about which products tend to be purchased together). The market basket analysis is not limited to the retail marketing domain and can be abstracted to other areas such as word co-occurrence relationships in text documents.
MultiPlot	Use the MultiPlot node to visualize data from a wide range of perspectives. The MultiPlot node automatically creates bar charts and scatter plots for the input and target variables without requiring you to make several menu or window item selections.
Path Analysis	Use the Path Analysis node to analyze Web log data and to determine the paths that visitors take as they navigate through a Web site. You can also use the node to perform sequence analysis.
SOM/Kohonen	Use the SOM/Kohonen node to perform unsupervised learning by using Kohonen vector quantization (VQ), Kohonen self-organizing maps (SOMs), or batch SOMs with Nadaraya-Watson or local-linear smoothing. Kohonen VQ is a clustering method, whereas SOMs are primarily dimension-reduction methods.
StatExplore	Use the StatExplore node to examine the statistical properties of an input data set. You can use the StatExplore node to compute standard univariate distribution statistics, to compute standard bivariate statistics by class target and class segment, and to compute correlation statistics for interval variables by interval input and target.
Variable Clustering	Variable clustering is a useful tool for data reduction and can remove collinearity, decrease variable redundancy, and help to reveal the underlying structure of the input variables in a data set. When properly used as a variable-reduction tool, the Variable Clustering node can replace a large set of variables with the set of cluster components with little loss of information.
Variable Selection	Use the Variable Selection node to quickly identify input variables that are useful for predicting the target variable.

Table A1.3 *Modify Nodes*

Node Name	Description
Drop	Use the Drop node to remove variables from data sets or hide variables from the metadata. You can drop specific variables and all variables of a particular type.
Impute	Use the Impute node to replace missing values. For example, you could replace missing values of an interval variable with the mean or using an M-estimator such as Andrew's Wave. Missing values for the training, validation, test, and score data sets are replaced using imputation statistics that are calculated from the active training predecessor data set.

Node Name	Description
Interactive Binning	The Interactive Binning node is an interactive grouping tool that you use to model nonlinear functions of multiple modes of continuous distributions. The interactive tool computes initial bins by quantiles. Then you can interactively split and combine the initial bins. This node enables you to select strong characteristics based on the Gini statistic and to group the selected characteristics based on business considerations. The node is helpful in shaping the data to represent risk-ranking trends rather than modeling quirks, which might lead to overfitting.
Principal Components	Use the Principal Components node to generate principal components, which are uncorrelated linear combinations of the original input variables and which depend on the covariance matrix or correlation matrix of the input variables. In data mining, principal components are usually used as the new set of input variables for subsequent analysis by modeling nodes.
Replacement	Use the Replacement node to generate score code to process unknown levels when scoring and also to interactively specify replacement values for class and interval levels. In some cases, you might want to reassign specified nonmissing values before performing imputation calculations for the missing values.
Rules Builder	The Rules Builder node accesses the Rules Builder window so you can create ad hoc sets of rules with user-definable outcomes. You can interactively define the values of the outcome variable and the paths to the outcome. This is useful, for example, in applying logic for posterior probabilities and scorecard values. Rules are defined using charts and histograms based on a sample of the data.
Transform Variables	Use the Transform Variables node to create new variables or variables that are transformations of existing variables in the data. Transformations are useful when you want to improve the fit of a model to the data. For example, transformations can be used to stabilize variances, remove nonlinearity, improve additivity, and correct non-normality in variables. The Transform Variables node also enables you to create interaction variables.

Table A1.4 *Model Nodes*

Node Name	Description
AutoNeural	Use the AutoNeural node as an automated tool to help you find optimal configurations for a neural network model.
Decision Tree	Use the Decision Tree node to fit decision tree models to the data. The implementation includes features that are found in a variety of popular decision tree algorithms such as CHAID, CART, and C4.5. The node supports both automatic and interactive training. When you run the Decision Tree node in automatic mode, it automatically ranks the input variables, based on the strength of their contribution to the tree. This ranking can be used to select variables for use in subsequent modeling. You can override any automatic step with the option to define a splitting rule and prune explicit tools or subtrees. Interactive training enables you to explore and evaluate a large set of trees as you develop them.

Node Name	Description
Dmine Regression	Use the Dmine Regression node to compute a forward stepwise least squares regression model. In each step, an independent variable is selected that contributes maximally to the model R-square value.
DMNeural	Use DMNeural node to fit an additive nonlinear model. The additive nonlinear model uses bucketed principal components as inputs to predict a binary or an interval target variable. The algorithm that is used in DMNeural network training was developed to overcome the problems of the common neural networks that are likely to occur especially when the data set contains highly collinear variables.
Ensemble	Use the Ensemble node to create new models by combining the posterior probabilities (for class targets) or the predicted values (for interval targets) from multiple predecessor models. One common ensemble approach is to use multiple modeling methods, such as a neural network and a decision tree, to obtain separate models from the same training data set. The component models from the two complementary modeling methods are integrated by the Ensemble node to form the final model solution
Gradient Boosting	Gradient boosting creates a series of simple decision trees that together form a single predictive model. Each tree in the series is fit to the residual of the prediction from the earlier trees in the series. Each time the data is used to grow a tree, the accuracy of the tree is computed. The successive samples are adjusted to accommodate previously computed inaccuracies. Because each successive sample is weighted according to the classification accuracy of previous models, this approach is sometimes called stochastic gradient boosting. Boosting is defined for binary, nominal, and interval targets.
LARS (Least Angle Regressions)	The LARs node can perform both variable selection and model-fitting tasks. When used for variable selection, the LARs node selects variables in a continuous fashion, where coefficients for each selected variable grow from zero to the variable's least square estimates. With a small modification, you can use LARs to efficiently produce LASSO solutions.
MBR (Memory-Based Reasoning)	Use the MBR node to identify similar cases and to apply information that is obtained from these cases to a new record. The MBR node uses k-nearest neighbor algorithms to categorize or predict observations.
Model Import	Use the Model Import node to import and assess a model that was not created by one of the SAS Enterprise Miner modeling nodes. You can then use the Model Comparison node to compare the user-defined model with one or more models that you developed with a SAS Enterprise Miner modeling node. This process is called integrated assessment.
Neural Network	Use the Neural Network node to construct, train, and validate multilayer, feed-forward neural networks. By default, the Neural Network node automatically constructs a network that has one hidden layer consisting of three neurons. In general, each input is fully connected to the first hidden layer, each hidden layer is fully connected to the next hidden layer, and the last hidden layer is fully connected to the output. The Neural Network node supports many variations of this general form.

Node Name	Description
Partial Least Squares	The Partial Least Squares node is a tool for modeling continuous and binary targets. This node extracts factors called components or latent vectors that can be used to explain response variation or predictor variation in the analyzed data.
Regression	Use the Regression node to fit both linear and logistic regression models to the data. You can use continuous, ordinal, and binary target variables, and you can use both continuous and discrete input variables. The node supports the stepwise, forward, and backward selection methods.
Rule Induction	Use the Rule Induction node to improve the classification of rare events. The Rule Induction node creates a Rule Induction model that uses split techniques to remove the largest pure split node from the data. Rule Induction also creates binary models for each level of a target variable and ranks the levels from the most rare event to the most common. After all levels of the target variable are modeled, the score code is combined into a SAS DATA step.
SVM (Support Vector Machines)	A support vector machine (SVM) is a supervised machine learning method that is used to perform classification and regression analysis. The standard SVM problem solves binary classification problems that produce non-probability output (only sign +1/-1) by constructing a set of hyperplanes that maximize the margin between two classes.
TwoStage	Use the TwoStage node to build a sequential or concurrent two-stage model for predicting a class variable and an interval target variable at the same time. The interval target variable is usually a value that is associated with a level of the class target.

Note: These modeling nodes use a directory table facility, called the Model Manager, in which you can store and access models on demand.

Table A1.5 *Assess Nodes*

Node Name	Description
Cutoff	The Cutoff node provides tabular and graphical information to assist you in determining an appropriate probability cutoff point for decision making with binary target models. The establishment of a cutoff decision point entails the risk of generating false positives and false negatives, but an appropriate use of the Cutoff node can help minimize those risks. You typically run the node at least twice. In the first run, you obtain all the plots and tables. In subsequent runs, you can change the node properties until an optimal cutoff value is obtained.
Decisions	Use the Decisions node to define target profiles for a target that produces optimal decisions. The decisions are made using a user-specified decision matrix and output from a subsequent modeling procedure.
Model Comparison	Use the Model Comparison node to compare models and predictions from any of the modeling nodes. The comparison is based on the expected and actual profits or losses that would result from implementing the model. The node produces the charts that help to describe the usefulness of the model.

Node Name	Description
Score	Use the Score node to manage SAS scoring code that is generated from a trained model or models, to save the SAS scoring code to a location on the client computer, and to run the SAS scoring code. Scoring is the generation of predicted values for a data set that might not contain a target variable.
Segment Profile	Use the Segment Profile node to examine segmented or clustered data and identify factors that differentiate data segments from the population. The node generates various reports that aid in exploring and comparing the distribution of these factors within the segments and population.

Table A1.6   *Utility Nodes*

Node Name	Description
Control Point	Use the Control Point node to establish a control point within process flow diagrams. A control point simplifies distributing the connections between process flow steps that have multiple interconnected nodes. The Control Point node can reduce the number of connections that are made.
End Groups	The End Groups node is used only in conjunction with the Start Groups node. The End Groups node acts as a boundary marker that defines the end of group processing operations in a process flow diagram. Group processing operations are performed on the portion of the process flow diagram that exists between the Start Groups node and the End Groups node. If you specify **Stratified**, **Bagging**, or **Boosting** in the group processing function of the Start Groups node,then the End Groups node functions as a model node and presents the final aggregated model.
Ext Demo	The Ext Demo node illustrates the various controls that can be used in SAS Enterprise Miner extension nodes. These controls enable users to pass arguments to an underlying SAS program. By choosing an appropriate user interface control, an extension node developer can specify how information about the node's arguments are presented to the user and place restrictions on the values of the arguments. The Ext Demo node's results also provide examples of the various types of graphs that can be generated by an extension node using the %EM_REPORT macro.
Metadata	Use the Metadata node to modify the columns metadata information (such as roles, measurement levels, and order) in a process flow diagram.
Reporter	The Reporter node uses SAS Output Delivery System (ODS) capability to create a single PDF or RTF file that contains information about the open process flow diagram. The report shows the SAS Enterprise Miner settings, process flow diagram, and detailed information for each node. The report also includes results such as variable selection, model diagnostic tables, and plots from the Results browser. The score code, log, and output listing are not included in the report; those items are found in the SAS Enterprise Miner package folder.
SAS Code	Use the SAS Code node to incorporate new or existing SAS code into process flows that you develop using SAS Enterprise Miner.

Node Name	Description
Score Code Export	The Score Code Export node is an extension for SAS Enterprise Miner that exports files that are necessary for score code deployment. Extensions are programmable add-ins for the SAS Enterprise Miner environment.
Start Groups	The Start Groups node is useful when the data can be segmented or grouped, and you want to process the grouped data in different ways. The Start Groups node uses BY-group processing as a method to process observations from one or more data sources that are grouped or ordered by values of one or more common variables.

Table A1.7   *Applications Nodes*

Node Name	Description
Incremetal Response	The Incremental Response node models the incremental impact of a treatment in order to optimize customer targeting for maximum return on investment. The Incremental Response node can determine the likelihood that a customer purchases a product or uses a coupon. It can predict the incremental revenue that is realized during a promotional period.
Ratemaking	Ratemaking is the process of determining which rates, or premiums, to charge each customer for their insurance. Traditional ratemaking methods are statistically unsophisticated, but the Ratemaking node uses generalized linear models, a proven technique, to analyze data and create a ratemaking model.
Survival	Survival data mining is the application of survival analysis to data mining problems concerning customers. The application to the business problem changes the nature of the statistical techniques. The issue in survival data mining is not whether an event will occur in a certain time interval, but when will the next event occurs. The SAS Enterprise Miner Survival node performs survival analysis on mining customer databases when there are time-dependent outcomes.

Usage Rules for Nodes

Here are some general rules that govern the placement of nodes in a process flow diagram:

- The Input Data Source node cannot be preceded by any other node.

- All nodes except the Input Data Source and SAS Code nodes must be preceded by a node that exports a data set.

- The SAS Code node can be defined in any stage of the process flow diagram. It does not require an input data set that is defined in the Input Data Source node.

- The Model Comparison node must be preceded by one or more modeling nodes.

- The Score node must be preceded by a node that produces score code. For example, the modeling nodes produce score code.

- The Ensemble node must be preceded by a modeling node.

- The Replacement node must follow a node that exports a data set, such as a Data Source, Sample, or Data Partition node.

Appendix 2
Sample Data Reference

The following table lists the variables that are included in the sample data sets donor_raw_data.sas7bdat and donor_score_data.sas7bdat.

Table A2.1 *Variables in the Sample Data Sets*

Variable	Description
CARD_PROM_12	number of card promotions sent to the individual by the charitable organization in the last 12 months
CLUSTER_CODE	one of 54 possible cluster codes, which are unique in terms of socioeconomic status, urbanicity, ethnicity, and other demographic characteristics
CONTROL_NUMBER	unique identifier of each individual
DONOR_AGE	age as of last year's mail solicitation
DONOR_GENDER	actual or inferred gender
FILE_AVG_GIFT	this variable is identical to LIFETIME_AVG_GIFT_AMT
FILE_CARD_GIFT	lifetime average donation (in $) from the individual in response to all card solicitations from the charitable organization
FREQUENCY_STATUS_97NK	based on the period of recency (determined by RECENCY_STATUS_96NK), which is the last 12 months for all groups except L and E, which are 13–24 months ago and 25–36 months ago, respectively: 1 if one donation in this period, 2 if two donations in this period, 3 if three donations in this period, 4 if four or more donations in this period
HOME_OWNER	H if the individual is a homeowner, U if this information is unknown
INCOME_GROUP	one of 7 possible income level groups based on a number of demographic characteristics

Variable	Description
IN_HOUSE	1 if the individual has ever donated to the charitable organization's In House program, 0 if not
LAST_GIFT_AMT	amount of the most recent donation from the individual to the charitable organization
LIFETIME_AVG_GIFT_AMT	lifetime average donation (in $) from the individual to the charitable organization
LIFETIME_CARD_PROM	total number of card promotions sent to the individual by the charitable organization
LIFETIME_GIFT_AMOUNT	total lifetime donation amount (in $) from the individual to the charitable organization
LIFETIME_GIFT_COUNT	total number of donations from the individual to the charitable organization
LIFETIME_GIFT_RANGE	maximum donation amount from the individual minus minimum donation amount from the individual
LIFETIME_MAX_GIFT_AMT	maximum donation amount (in $) from the individual to the charitable organization
LIFETIME_MIN_GIFT_AMT	minimum donation amount (in $) from the individual to the charitable organization
LIFETIME_PROM	total number of promotions sent to the individual by the charitable organization
MEDIAN_HOME_VALUE	median home value (in $100) as determined by other input variables
MEDIAN_HOUSEHOLD_INCOME	median household income (in $100) as determined by other input variables
MONTHS_SINCE_FIRST_GIFT	number of months since the first donation from the individual to the charitable organization
MONTHS_SINCE_LAST_GIFT	number of months since the most recent donation from the individual to the charitable organization
MONTHS_SINCE_LAST_PROM_RESP	number of months since the individual has responded to a promotion by the charitable organization
MONTHS_SINCE_ORIGIN	number of months that the individual has been in the charitable organization's database

Variable	Description
MOR_HIT_RATE	total number of known times the donor has responded to a mailed solicitation from a group other than the charitable organization
NUMBER_PROM_12	number of promotions (card or other) sent to the individual by the charitable organization in the last 12 months
OVERLAY_SOURCE	the data source against which the individual was matched: M if Metromail, P if Polk, B if both
PCT_ATTRIBUTE1	percent of residents in the neighborhood in which the individual lives that are males and active military
PCT_ATTRIBUTE2	percent of residents in the neighborhood in which the individual lives that are males and veterans
PCT_ATTRIBUTE3	percent of residents in the neighborhood in which the individual lives that are Vietnam veterans
PCT_ATTRIBUTE4	percent of residents in the neighborhood in which the individual lives that are WWII veterans
PCT_OWNER_OCCUPIED	percent of owner-occupied housing in the neighborhood in which the individual lives
PEP_STAR	1 if individual has ever achieved STAR donor status, 0 if not
PER_CAPITA_INCOME	per capita income (in $) of the neighborhood in which the individual lives
PUBLISHED_PHONE	1 if the individual's telephone number is published, 0 if not
RECENCY_STATUS_96NK	recency status as of two years ago: A if active donor, S if star donor, N if new donor, E if inactive donor, F if first time donor, L if lapsing donor
RECENT_AVG_CARD_GIFT_AMT	average donation from the individual in response to a card solicitation from the charitable organization since four years ago
RECENT_AVG_GIFT_AMT	average donation (in $) from the individual to the charitable organization since four years ago

Variable	Description
RECENT_CARD_RESPONSE_COUNT	number of times the individual has responded to a card solicitation from the charitable organization since four years ago
RECENT_CARD_RESPONSE_PROP	proportion of responses to the individual to the number of card solicitations from the charitable organization since four years ago
RECENT_RESPONSE_COUNT	number of times the individual has responded to a promotion (card or other) from the charitable organization since four years ago
RECENT_RESPONSE_PROP	proportion of responses to the individual to the number of (card or other) solicitations from the charitable organization since four years ago
RECENT_STAR_STATUS	1 if individual has achieved star donor status since four years ago, 0 if not
SES	one of 5 possible socioeconomic codes classifying the neighborhood in which the individual lives
TARGET_B	1 if individual donated in response to last year's 97NK mail solicitation from the charitable organization, 0 if individual did not
TARGET_D	amount of donation (in $) from the individual in response to last year's 97NK mail solicitation from the charitable organization
URBANICITY	classification of the neighborhood in which the individual lives: U if urban, C if city, S if suburban, T if town, R if rural, ? if missing
WEALTH_RATING	one of 10 possible wealth rating groups based on a number of demographic characteristics

Glossary

assessment
the process of determining how well a model computes good outputs from input data that is not used during training. Assessment statistics are automatically computed when you train a model with a modeling node. By default, assessment statistics are calculated from the validation data set.

champion model
the best predictive model that is chosen from a pool of candidate models in a data mining environment. Candidate models are developed using various data mining heuristics and algorithm configurations. Competing models are compared and assessed using criteria such as training, validation, and test data fit and model score comparisons.

chi-squared automatic interaction detection
a technique for building decision trees. The CHAID technique specifies a significance level of a chi-square test to stop tree growth. Short-form: CHAID.

classification and regression trees
a decision tree technique that is used for classifying or segmenting a data set. The technique provides a set of rules that can be applied to new data sets in order to predict which records will have a particular outcome. It also segments a data set by creating 2-way splits. The CART technique requires less data preparation than CHAID. Short form: CART.

data mining database
a SAS data set that is designed to optimize the performance of the modeling nodes. DMDBs enhance performance by reducing the number of passes that the analytical engine needs to make through the data. Each DMDB contains a meta catalog, which includes summary statistics for numeric variables and factor-level information for categorical variables. Short form: DMDB.

data source
a data object that represents a SAS data set in the Java-based Enterprise Miner GUI. A data source contains all the metadata for a SAS data set that Enterprise Miner needs in order to use the data set in a data mining process flow diagram. The SAS data set metadata that is required to create an Enterprise Miner data source includes the name and location of the data set, the SAS code that is used to define its library path, and the variable roles, measurement levels, and associated attributes that are used in the data mining process.

decision tree

the complete set of rules that are used to split data into a hierarchy of successive segments. A tree consists of branches and leaves, in which each set of leaves represents an optimal segmentation of the branches above them according to a statistical measure.

dependent variable

a variable whose value is determined by the value of another variable or by the values of a set of variables.

depth

the number of successive hierarchical partitions of the data in a tree. The initial, undivided segment has a depth of 0.

Gini index

a measure of the total leaf impurity in a decision tree.

hidden layer

in a neural network, a layer between input and output to which one or more activation functions are applied. Hidden layers are typically used to introduce nonlinearity.

imputation

the computation of replacement values for missing input values.

input variable

a variable that is used in a data mining process to predict the value of one or more target variables.

interval variable

a continuous variable that contains values across a range. For example, a continuous variable called Temperature could have values such as 0, 32, 34, 36, 43.5, 44, 56, 80, 99, 99.9, and 100.

leaf

in a tree diagram, any segment that is not further segmented. The final leaves in a tree are called terminal nodes.

logistic regression

a form of regression analysis in which the target variable (response variable) represents a binary-level or ordinal-level response.

metadata

a description or definition of data or information.

MLP

See multilayer perceptron.

model

a formula or algorithm that computes outputs from inputs. A data mining model includes information about the conditional distribution of the target variables, given the input variables.

multilayer perceptron

a neural network that has one or more hidden layers, each of which has a linear combination function and executes a nonlinear activation function on the input to that layer. Short form: MLP. See also hidden layer.

neural networks

a class of flexible nonlinear regression models, discriminant models, data reduction models, and nonlinear dynamic systems that often consist of a large number of neurons. These neurons are usually interconnected in complex ways and are often organized into layers. See also neuron.

node

(1) in the SAS Enterprise Miner user interface, a graphical object that represents a data mining task in a process flow diagram. The statistical tools that perform the data mining tasks are called nodes when they are placed on a data mining process flow diagram. Each node performs a mathematical or graphical operation as a component of an analytical and predictive data model. (2) in a neural network, a linear or nonlinear computing element that accepts one or more inputs, computes a function of the inputs, and can direct the result to one or more other neurons. Nodes are also known as neurons or units. (3) a leaf in a tree diagram. The terms leaf, node, and segment are closely related and sometimes refer to the same part of a tree. See also process flow diagram and internal node.

observation

a row in a SAS data set. All of the data values in an observation are associated with a single entity such as a customer or a state. Each observation contains either one data value or a missing-value indicator for each variable.

overfit

to train a model to the random variation in the sample data. Overfitted models contain too many parameters (weights), and they do not generalize well. See also underfit.

partition

to divide available data into training, validation, and test data sets.

PFD

See process flow diagram.

predicted value

in a regression model, the value of a dependent variable that is calculated by evaluating the estimated regression equation for a specified set of values of the explanatory variables.

prior probability

a probability that reflects knowledge about the population before obtaining the sample on hand.

process flow diagram

a graphical representation of the various data mining tasks that are performed by individual Enterprise Miner nodes during a data mining analysis. A process flow diagram consists of two or more individual nodes that are connected in the order in which the data miner wants the corresponding statistical operations to be performed. Short form: PFD.

profit matrix

a table of expected revenues and expected costs for each decision alternative for each level of a target variable.

project

a user-created GUI entity that contains the related SAS Enterprise Miner components required for the data mining models. A project contains SAS Enterprise Miner data sources, process flow diagrams, and results data sets and model packages.

pruning

the process of removing nodes from a decision tree when those nodes involve less than optimal decision rules.

root node

the initial segment of a tree. The root node represents the entire data set that is submitted to the tree, before any splits are made.

sampling

the process of subsetting a population into n cases. Sampling decreases the time required for fitting a model.

SAS data set

a file whose contents are in one of the native SAS file formats. There are two types of SAS data sets: SAS data files and SAS data views. SAS data files contain data values in addition to descriptor information that is associated with the data. SAS data views contain only the descriptor information plus other information that is required for retrieving data values from other SAS data sets or from files that are stored in other software vendors' file formats.

scoring

the process of applying a model to new data in order to compute outputs. Scoring is the last process that is performed in data mining.

SEMMA

the data mining process that is used by Enterprise Miner. SEMMA stands for Sample, Explore, Modify, Model, and Assess.

subdiagram

in a process flow diagram, a collection of nodes that are compressed into a single node. The use of subdiagrams can improve your control of the information flow in the diagram.

target variable

a variable whose values are known in one or more data sets that are available (in training data, for example) but whose values are unknown in one or more future data sets (in a score data set, for example). Data mining models use data from known variables to predict the values of target variables.

training

the process of computing good values for the weights in a model.

training data

currently available data that contains input values and target values that are used for model training.

transformation

the process of applying a function to a variable in order to adjust the variable's range, variability, or both.

tree

the complete set of rules that are used to split data into a hierarchy of successive segments. A tree consists of branches and leaves, in which each set of leaves represents an optimal segmentation of the branches above them according to a statistical measure.

tree structure

a type of data structure that uses the graphic analogy of a tree with branches and leaves. Each set of leaves represents an optimal segmentation of the branches above it, according to a statistical measure and the rules that govern the structure.

underfit

to train a model to only part of the actual patterns in the sample data. Underfit models contain too few parameters (weights), and they do not generalize well. See also overfit.

validation data

data that is used to validate the suitability of a data model that was developed using training data. Both training data sets and validation data sets contain target variable values. Target variable values in the training data are used to train the model. Target variable values in the validation data set are used to compare the training model's predictions to the known target values, assessing the model's fit before using the model to score new data.

variable

a column in a SAS data set or in a SAS data view. The data values for each variable describe a single characteristic for all observations. Each SAS variable can have the following attributes: name, data type (character or numeric), length, format, informat, and label.

variable attribute

any of the following characteristics that are associated with a particular variable: name, label, format, informat, data type, and length.

variable level

the set of data dimensions for binary, interval, or class variables. Binary variables have two levels. A binary variable CREDIT could have levels of 1 and 0, Yes and No, or Accept and Reject. Interval variables have levels that correspond to the number of interval variable partitions. For example, an interval variable PURCHASE_AGE might have levels of 0-18, 19-39, 40-65, and >65. Class variables have levels that correspond to the class members. For example, a class variable HOMEHEAT might have four variable levels: Coal/Wood, FuelOil, Gas, and Electric. Data mining decision and profit matrixes are composed of variable levels.

Index

CPSIA information can be obtained
at www.ICGtesting.com
Printed in the USA
LVOW09s2255210417
531776LV00006B/24/P